A Closer Approximation of ☉ On the MATW

Books by Jaime Jackson

Equine

The Natural Horse: Lessons from the Wild (1992, rev. 2020)
Horse Owners Guide to Natural Hoof Care (1999, rev. 2002)
Founder – Prevention and Cure the Natural Way (2001)
Guide To Booting Horses for Hoof Care Professionals (2002)
Paddock Paradise: A Guide to Natural Horse Boarding (2005, rev. 2018)
The Natural Trim: Principles and Practice (2012, rev. 2019)
The Healing Angle: Nature's Gateway to the Healing Field (2014)
Laminitis: An Equine Plague of Unconscionable Proportions (2016)
Training Manual: ISNHCP Natural Trim Training Program (2017)
the Hoof Balancer: A Unique Tool for Balancing Equine Hooves (2019)
The Natural Trim: Basic Guidelines (2019)
The Natural Trim: Advanced Guidelines (2019)
Navicular Syndrome: Healing And Prevention Using the Principles and Practices of Natural Horse Care (2021)
A Closer Approximation of ☉ On the MATW Using An Infrared Thermometer With Laser Pointer Gun (2021)

Other

Guard Your Teeth: Why the Dental Industry Fails Us – A Guide to Natural Dental Care (2018)
Buckskin Tanner: A Guide to Natural Hide Tanning (2019)
Cheyenne Tipi Notes: Technical Insights Into 19th Century Plains Indian Bison Hide Tanning (2019)
Living Behind the Facade: Memoirs Of A Gay Man's Journey Through the 20th Century (2019) George Somers with Jaime Jackson
Platform: A Humanitarian Model For An Egalitarian Society (2019)
Zoo Paradise: A New Model for Humane Zoological Gardens (2019)

Forthcoming
Horse Trek – Into the Mystic

A Closer Approximation of ☉ On the MATW
Using An Infrared Thermometer With Laser Pointer Gun

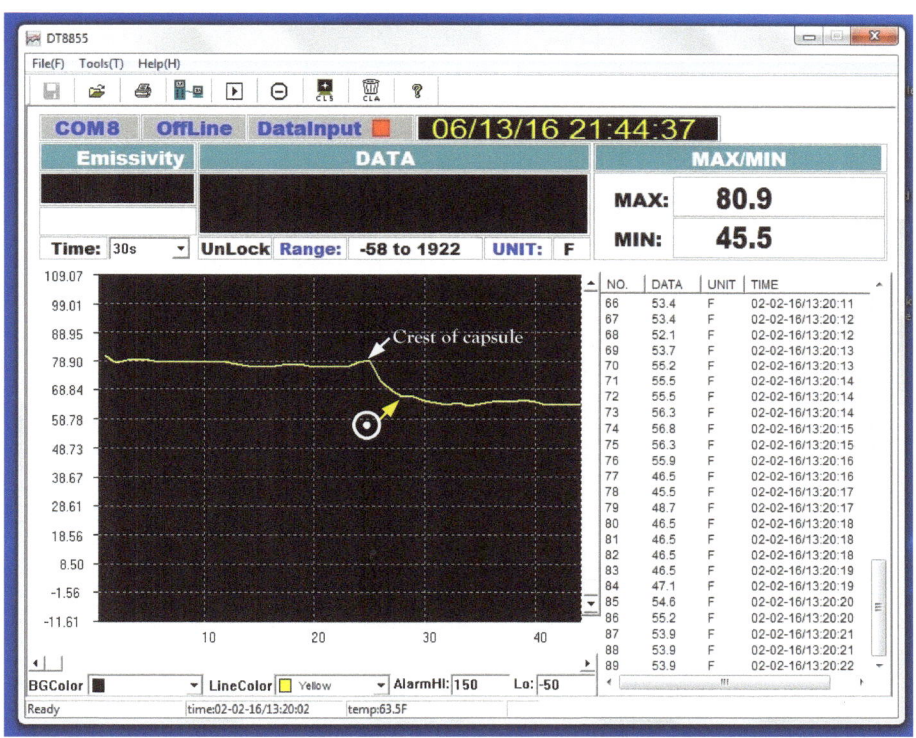

Jaime Jackson
Author, *The Natural Horse: Lessons From the Wild*

J. JACKSON PUBLISHING

Copyright © 2021 Jaime Jackson
All rights reserved.

This book may not be reproduced in whole or in part, by any means (with the exception of short quotes for the purpose of review), without permission of the publisher.

J. Jackson Publishing
P.O. Box 1765
Harrison, AR 72601
jjacksonpublishing@gmail.com

ISBN: 978-1-7355358-3-8

Manufactured in the United States of America by
Ingram Book Company (https://www.ingramcontent.com)

Neither the author, contributors nor J. Jackson Publishing will accept responsibility for any illness, injury, or death resulting from your use and/or application of this work. The ideas, procedures, and suggestions contained in this book are not meant as substitutes for consultation with your own veterinarian, hoof care provider, or other professional. You, the reader, are strongly urged to use your own judgment, intuition, and common sense in utilizing this book.

Contents

Introduction:	*1*
Using ☉ to determine H° and H°TL	*2*
Rationale for ☉	*4*
Pin-pointing ☉ using infrared thermal technology	*5*
Laser Scan of MATW	*8*
Laser Scanned Graph of MATW	*9*
Thermal Zones On the MATW	*10*
Using the Laser Gun	*11*
Measuring H° and H°TL · Measuring B° and B°TL	*12*
Impact of ambient temperatures on the location of ⊕	*14*
Impact of laminitis on thermal flucuations along the MATW	*15*
Definitions	*16*
Text References	*18*
Attributions	*18*
Study Technology	*18*
NHC Links	*19*

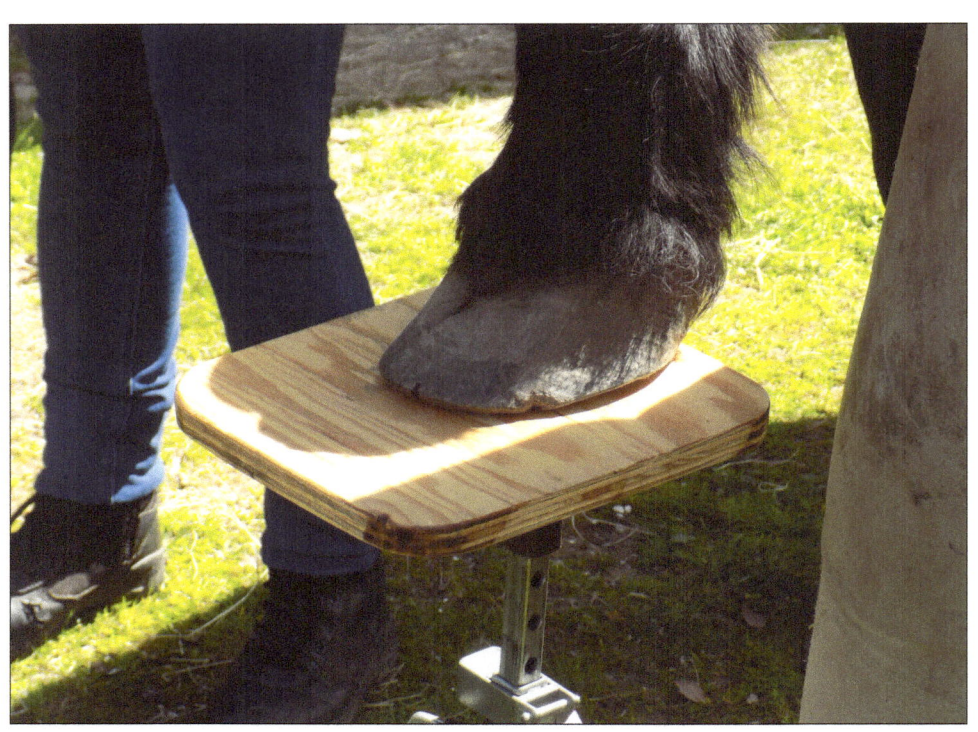

Introduction
A Closer Approximation of ☉ On the MATW
Using An Infrared Thermometer With Laser Pointer Gun

The location of ☉ (Bull's-eye) on the Median Axis of the Toe Wall (MATW) is a core Navigational Landmark in the Natural Trim Hoof Plexus (*Figure 1*).[1] According to the theory of H°, ☉ enables the Natural Hoof Care (NHC) Practitioner to find their way to both H° and H°TL. These two Critical Measurements are foundational to the Natural Trim as they correlate directly with their ancient wilderness counterparts, namely N° and N°TL, derived from America's wild, free-roaming horses.[2] Without these correlations linking the domesticated horse to the wild, free-roaming horse, there would be no genuine

---

[1] Jaime Jackson. *The Natural Trim: Basic Guidelines –Trim Mechanics, Biodynamics, and healing Forces in Paddock Paradise: Working with Naturre to Create the Pefect Hoof.* p. 51, Fig. 5-4.

[2] This adaptive environment of Equus caballus is exempled by the U.S. Great Basin. N° and N°TL were derived from America's wild horses living there by the author during his 1982-1986 studies at the Bureau of Land Management's wild horse processing corrals near Litchfield, California (USA). *The Natural Horse: Lessons From the Wild* (2020 ed.). p. 108, Fig. 6-1.

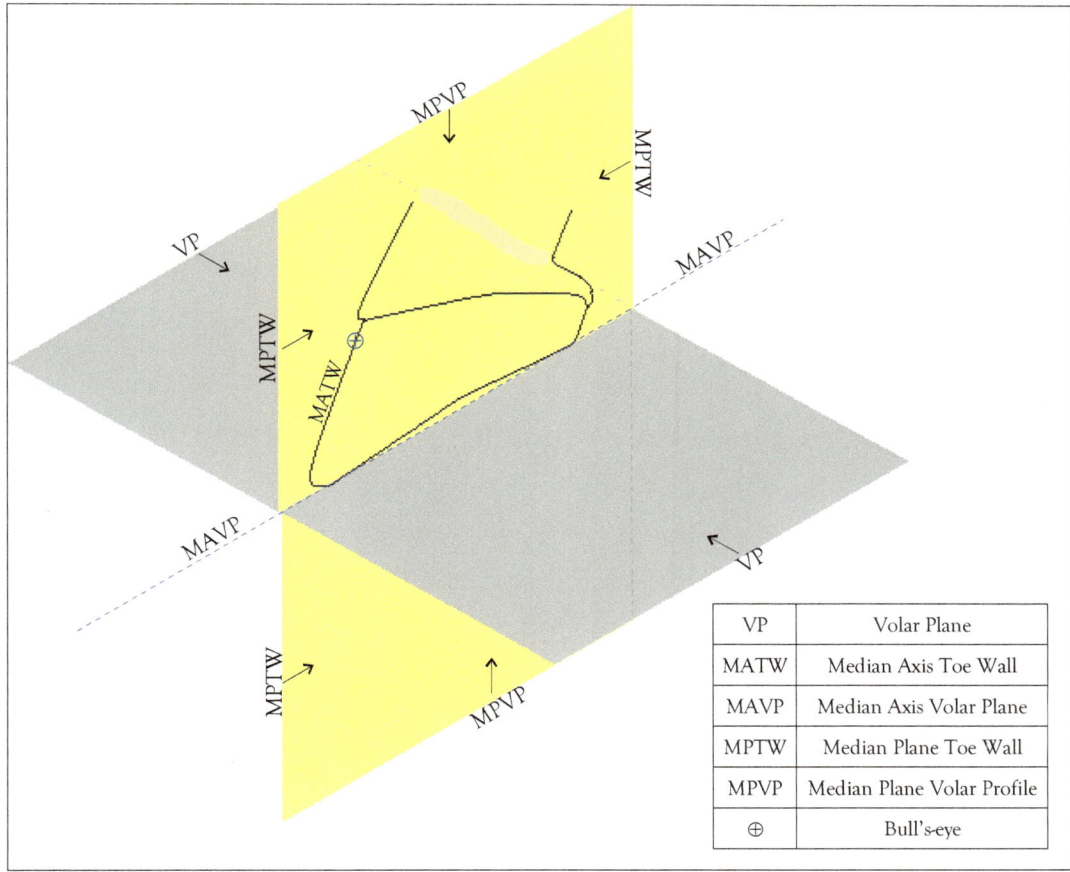

Figure 1. The *Hoof Plexus*.

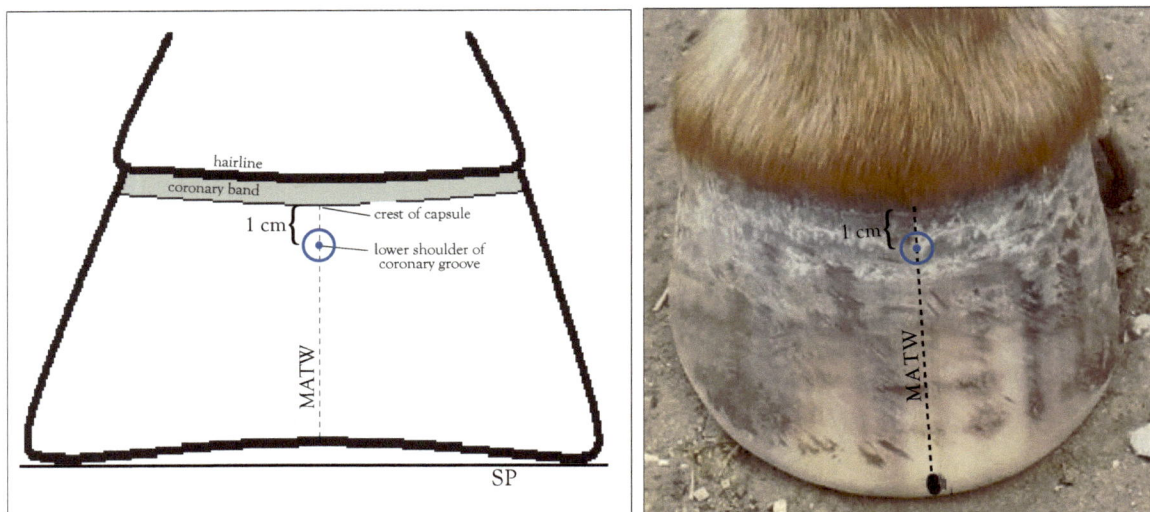

Figure 2. Location of ☉ 1 cm below crest of capsule on the MATW.

Natural Trim in the world today. Hoof care would then default to the crude hoof care theories and practices borne of medieval Europe.

☉ first appeared in the official NHC lexicon in the early days of the American Association of Natural Hoof Care Practices (AANHCP) dating back to 2000.[1] Initially, it was called the "limit line," a now defunct term, used in lectures before it arrived in print several years later. Eventually, as the language and understanding of hoof morphology evolved within the realm of NHC, ☉ arrived in print in different documents and in early versions of *The Natural Trim: Principles and Practice*.[2] In 2009 NHC training was transferred from the AANHCP to the Institute for the Study of Natural Horse Care Practices (ISNHCP), where ☉ has since been part of the core training.

Using ☉ to determine H° and H°TL

Technically, ☉ is currently defined as a single point on the MATW, approximately 1.0 cm below the *crest of the capsule* (*Figure 2*). This point, or dot drawn on the MATW, corresponds to the approximate location where the full thickness (width) of the hoof wall begins (*Figure 3*). Above (superior to) ☉, the hoof wall is extruded as a relatively thin rim of epidermis at its juncture with the perioplic groove (*Figure 4*). From there it tapers down to the lower shoulder of the

---

[1]Today, reflecting the organization's international membership and broad holistic approach to equine care, the acronym stands for Association for the Advancement of Natural Horse Care Practices.

[2]Jaime Jackson. *The Natural Trim: Principles and Practice* (2012; rev. 2019).

Using ☉ to determine H° and H°TL

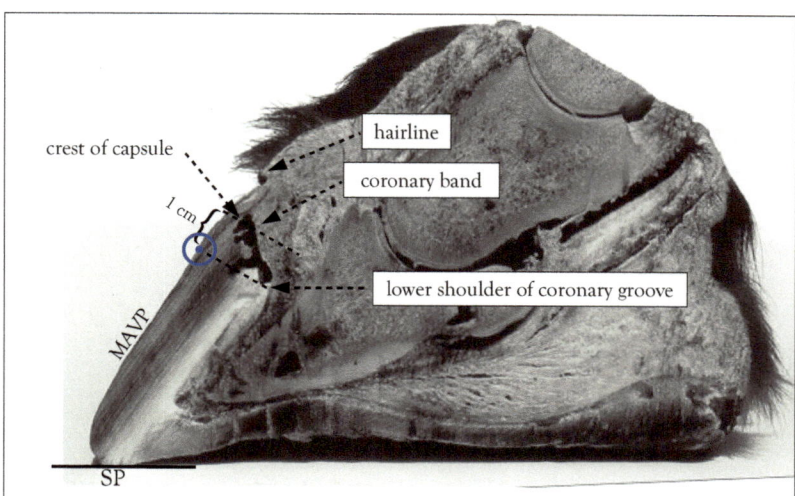

Figure 3. Location of ☉ 1 cm below crest of capsule on the MATW. A theoretical line drawn perpendicularly from ☉ such that it intersects the lower shoulder of the coronary groove marks the point at which the newly grown toe wall matures into its full width.

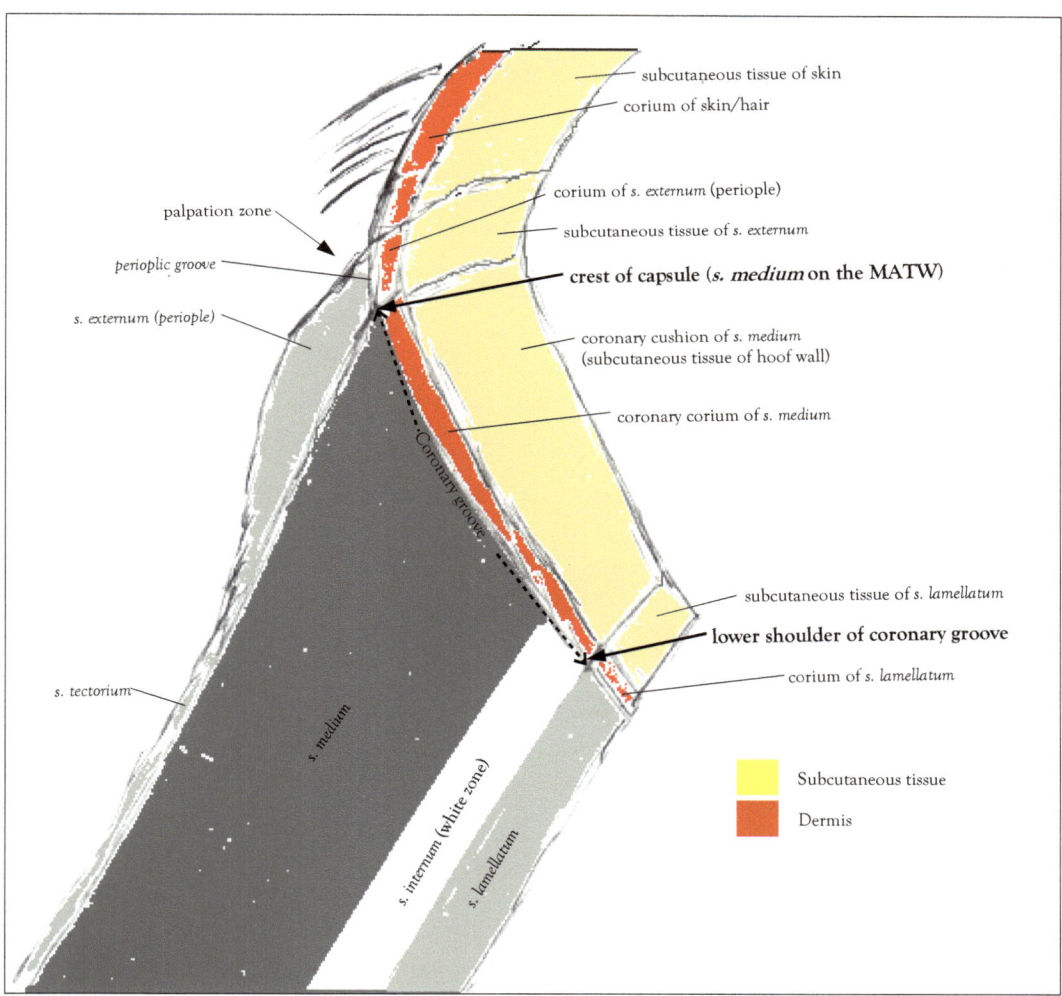

Figure 4. Structures of the hoof wall.

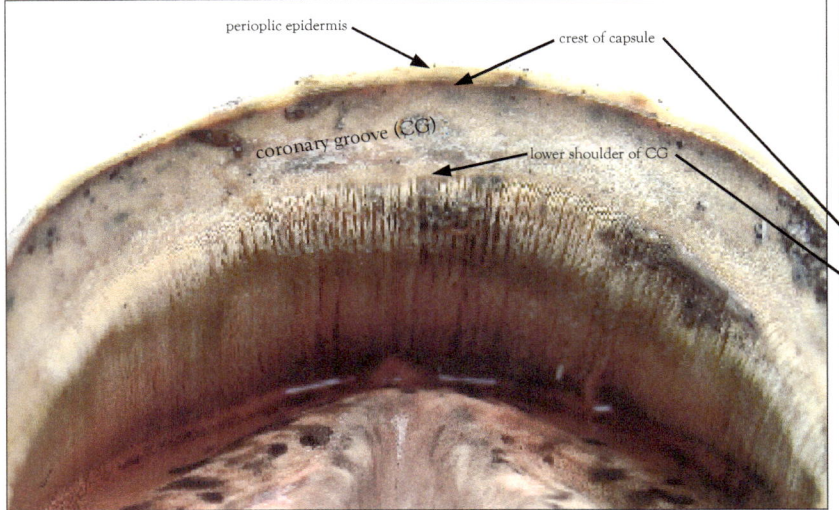

Figure 5. Relative positions of the crest of capsule, coronary groove, and lower shoulder of coronary groove.

coronary groove where its full thickness (width) is established (*Figure 5*). Natural Trim guidelines specify measuring H°TL from ☉ down to the hoof's Volar Plane (VP); VP is defined by the hoof's Support Plane (SP) (*Figure 6*).

Rationale for ☉

☉ arose as a Navigational Landmark as a way to bypass the natural accumulation of periople (perioplic epidermis) below the crest of the capsule. This epidermis obstructs the use of a straight edge ruler or the Hoof Meter

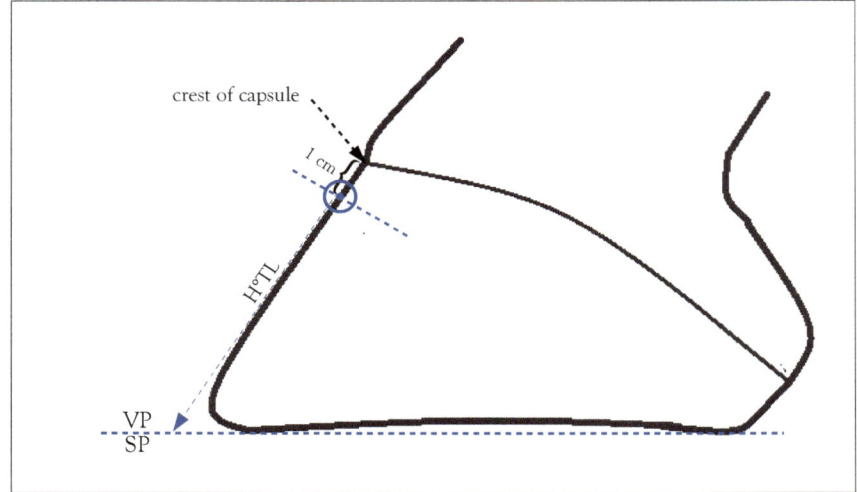

Figure 6. HTL is measured from ☉ to the hoof's Volar Plane (VP). VP is defined by the hoof's Support Plane (SP).

Reader (HMR) Toe Length Reader from laying flat against and parallel with the MATW. Rasping away the periople to facilitate such a straight-edge device was not an option either — nature intends periople to be there to cover and protect the coriums of the *s. externum* and the *s. medium* (*Figure 4*). Thus, the trimmer had no choice but to somehow bypass the perioplic epidermis altogether in order to measure H°TL.

From this dilemma — and in an attempt to bring further definition to the *theory of H°* (p.17) — I made the decision to measure H°TL from ⊙ to SP in order to bypass the perioplic obstacle, but also to hone in further on the point at which the full natural thickness of the hoof wall begins. Not surprisingly, I was immediately faced with yet another nagging question: If ⊙ is to define the location where the full thickness of the hoof wall begins, then where exactly is the lower shoulder of the coronary groove?

Pin-pointing ⊙ using infrared thermal technology

Many dissections of cadaver specimens are revealing that ⊙ may vary from 1.0 to 1.5 cm (and sometimes more) below the crest of the capsule. So precision is questionable at best, even if there was no periople to interfere with a ruler or the HMR. This uncertainty, as one can appreciate, challenges the trimmer to make anything more than a rough estimate of where ⊙ is located on the MATW.

In searching for a more precise way to pinpoint ⊙ on the MATW, my attention was brought to the vascular network of the *Supercorium* (p.17) including its contribution to the foot's *homeostasis* (p. 16). The foot's vascular system lies just behind the entire epidermal armor comprising the hoof. Thus, it seemed logical that thermal readings taken on the hoof wall along the MATW would fluctuate according to any given segment's relative proximity to the Supercorium. In this interpretation, I anticipated that there would be both "cool zones" and "warm zones" along the MATW.

With this premise in mind, I speculated that thermal readings on the MATW for °F (or °C) would vary from the crest of the capsule down to SP as follows (*Figure 7*):

- The coronary groove would be a *warm zone* due to the presence of the circumflex artery feeding the hoof wall's vast coronary corium (coronary band). And because it is naturally the thinnest epidermal region of the hoof wall —

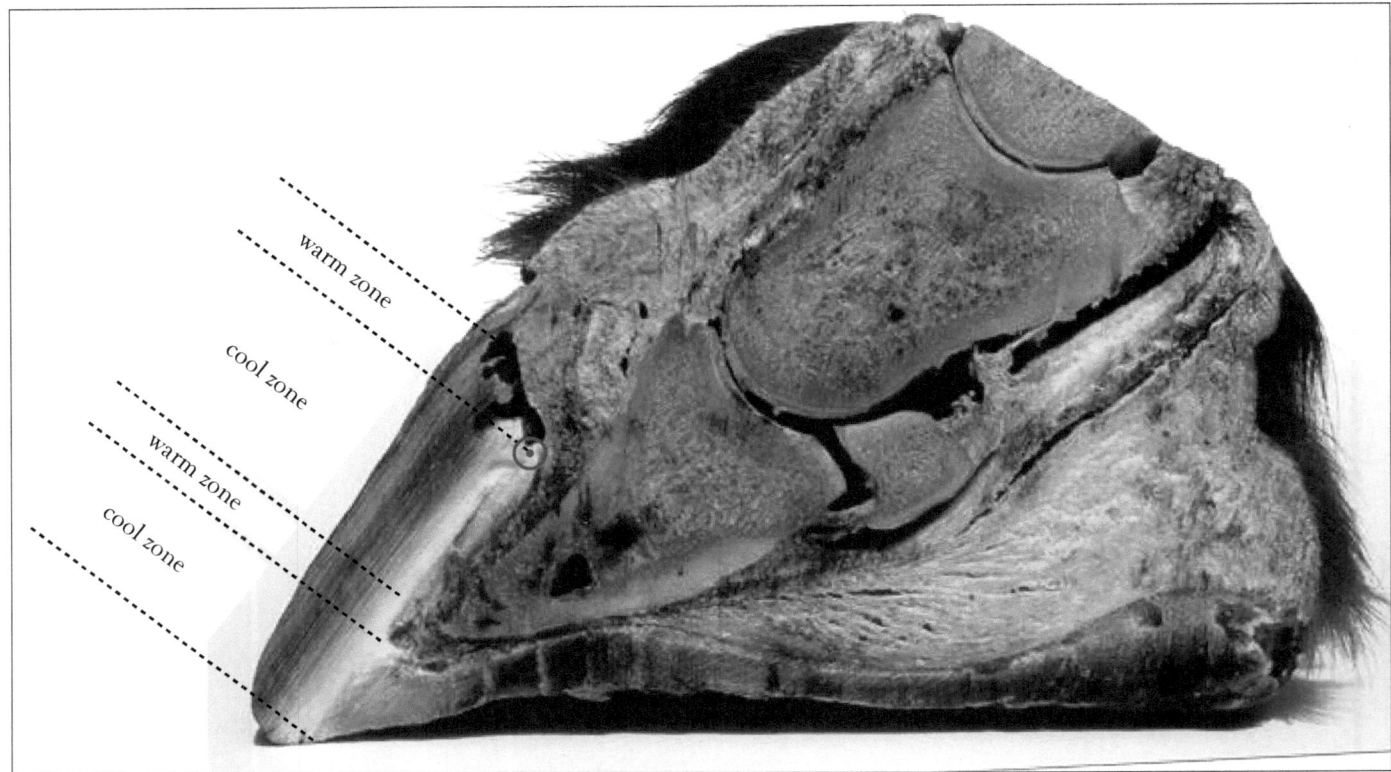

Figure 7. Projected "Hot" and "Cool" zones on the MATW.

until its lower shoulder completes the *stratum internum* (the "white zone"), at which point the full thickness of the hoof wall, excluding the epidermal *stratum lamellatum*, is established. ⊙ is located here, a cool point on the MATW, bridging the coronary groove with the inner hoof wall.

- Next to arrive would be a *cool zone* comprising the entire length of the hoof wall from its lower shoulder on the coronary groove, designated as ⊙, down to its juncture with the sole corium. Along this stretch, the inner toe wall's epidermal laminae interdigitate with the dermal laminae, forming a thin strip of vascular tissue.

- Further down, a second *warm zone* would occur along the MATW where the inner hoof wall interfaces with the lamellar and solar coriums to form the epidermal lamellar bridge between the hoof wall and the sole.

- A final *cool zone* begins where the inner hoof wall and epidermal sole come together, forming the epidermal *stratum lamellatum*, thereafter terminating at ground level.

If the foregoing were proven true, I concluded that ⊙ should occur precisely at the lowest point on the coronary groove (*Figure 7*). I decided to test this hypothesis with the hooves of our horses at the AANHCP Paddock Paradise near Lompoc, California (USA).

To gauge temperatures along the MATW, I deployed an infrared (IR) thermometer with a laser pointer gun connected via bluetooth to my laptop computer.[1]

As I anticipated, there was a distinct correlation between the termperature readings along the MATW and the epidermal structures comprising the hoof wall predicted in my hypothesis. Here are my findings for Apollo, one of the horses living 24/7 on track with four other horses.

- Figure 8 depicts the results of the laser scan in graphic representation as seen on my laptop.

- Figure 9 graphs the corresponding data flow in °F/t (Fahrenheit/seconds).

- Figure 10 tabulates the data flow in °F/t with time signature.

(Continued on page 11)

[1]Model: WCI Professional High-Temperature IR Thermometer Laser Pointer Gun with Wireless USB Computer Interface and Type K Probe - Instant °C or °F Measurement with Calibrated Graphics Software.

(Overleaf): Figures 8-10

Laser Scan of MATW

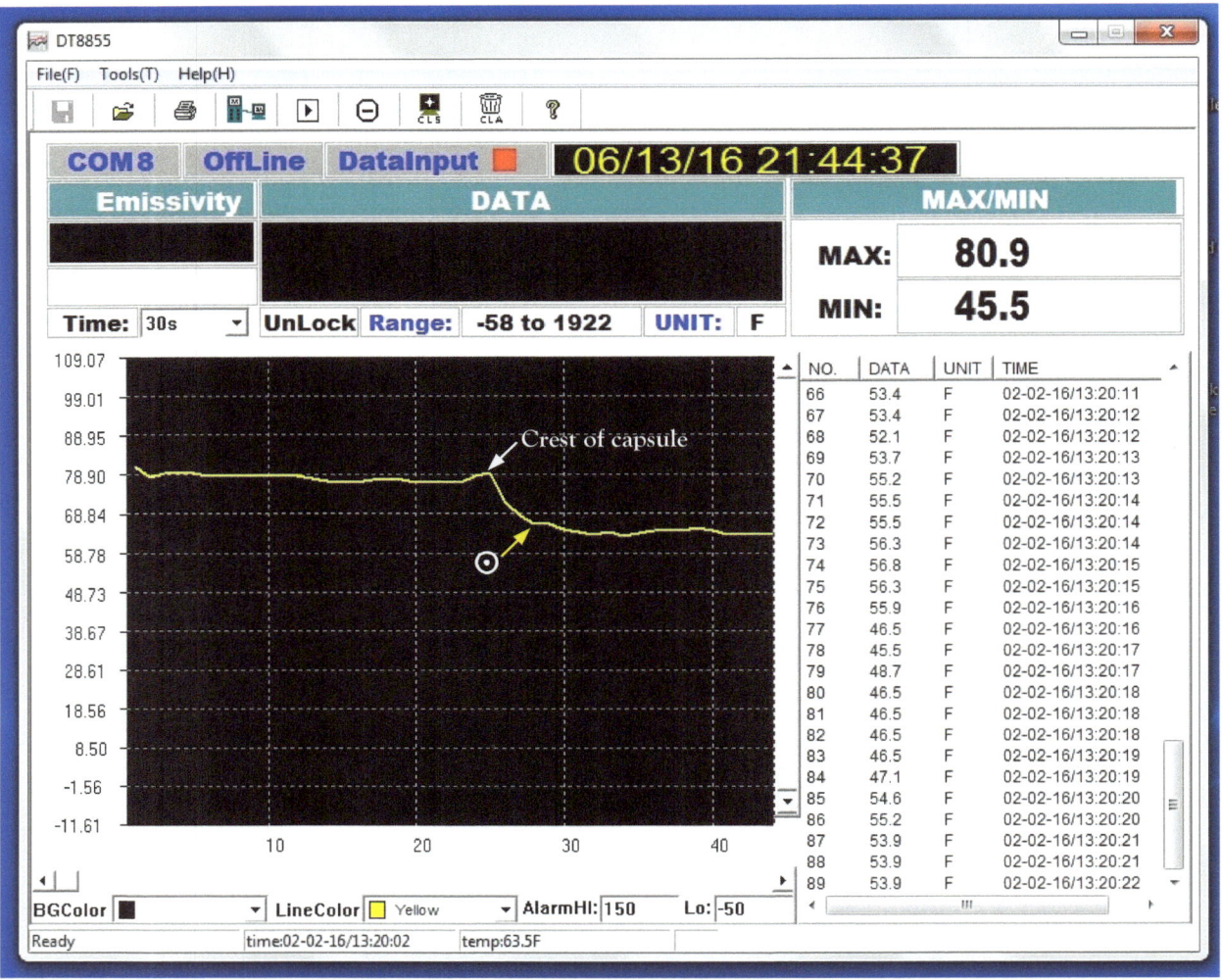

Figure 8: Laser Scan of MATW
- *White arrow* points to crest of capsule.
- *Yellow arrow* points to lower shoulder of coronary groove. ⊙ is located here.

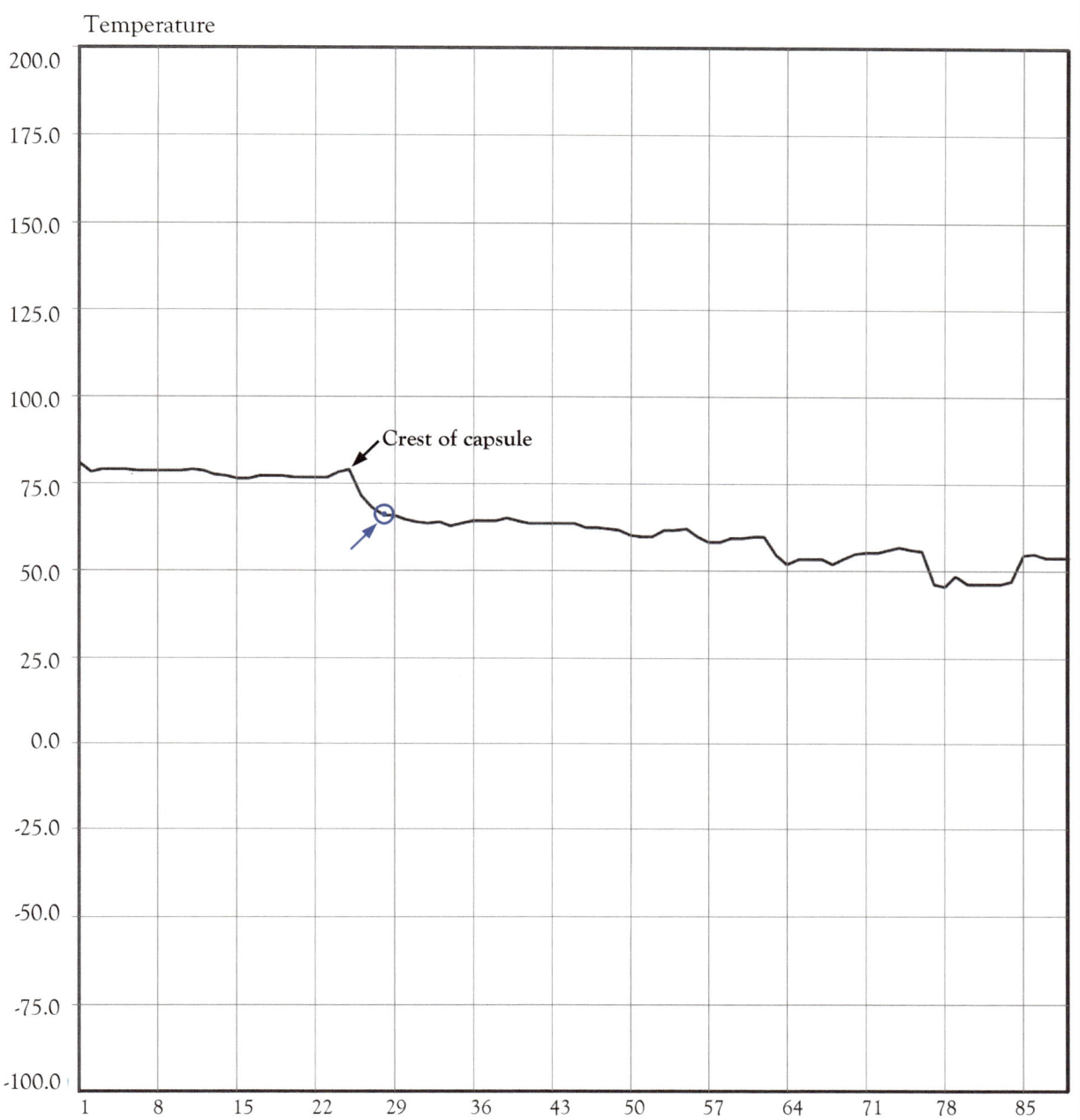

Figure 9: Laser Scanned Graph of MATW
- *Black arrow* points to crest of capsule.
- *Blue arrow* points to lower shoulder of coronary groove. ⊙ is located here.

Thermal Zones On the MATW

NO	DATA	UNIT	TIME
1	80.9	F	02-02-16/13:19:42
2	78.2	F	02-02-16/13:19:43
3	79.1	F	02-02-16/13:19:43
4	79.1	F	02-02-16/13:19:44
5	79.1	F	02-02-16/13:19:44
6	78.6	F	02-02-16/13:19:45
7	78.6	F	02-02-16/13:19:45
8	78.6	F	02-02-16/13:19:46
9	78.6	F	02-02-16/13:19:46
10	78.6	F	02-02-16/13:19:46
11	78.9	F	02-02-16/13:19:47
12	78.6	F	02-02-16/13:19:47
13	77.7	F	02-02-16/13:19:48
14	77.0	F	02-02-16/13:19:48
15	76.6	F	02-02-16/13:19:49
16	76.6	F	02-02-16/13:19:49
17	77.3	F	02-02-16/13:19:50
18	77.3	F	02-02-16/13:19:50
19	77.3	F	02-02-16/13:19:50
20	76.8	F	02-02-16/13:19:51
21	76.8	F	02-02-16/13:19:51
22	76.8	F	02-02-16/13:19:52
23	76.8	F	02-02-16/13:19:52
24	78.4	F	02-02-16/13:19:53
25	79.1	F	02-02-16/13:19:53
26	71.6	F	02-02-16/13:19:54
27	68.3	F	02-02-16/13:19:54
28	66.0 ☉	F	02-02-16/13:19:54
29	66.0	F	02-02-16/13:19:55
30	64.7	F	02-02-16/13:19:55
31	64.2	F	02-02-16/13:19:56
32	63.5	F	02-02-16/13:19:56
33	64.2	F	02-02-16/13:19:57
34	63.1	F	02-02-16/13:19:57
35	63.8	F	02-02-16/13:19:58
36	64.4	F	02-02-16/13:19:58
37	64.4	F	02-02-16/13:19:58
38	64.4	F	02-02-16/13:19:59
39	65.3	F	02-02-16/13:19:59
40	64.5	F	02-02-16/13:20:00
41	63.5	F	02-02-16/13:20:00
42	63.5	F	02-02-16/13:20:01
43	63.5	F	02-02-16/13:20:01
44	63.5	F	02-02-16/13:20:02
45	63.5	F	02-02-16/13:20:02
46	62.6	F	02-02-16/13:20:02
47	62.6	F	02-02-16/13:20:03
48	62.0	F	02-02-16/13:20:03
49	61.7	F	02-02-16/13:20:04
50	60.4	F	02-02-16/13:20:04
51	60.0	F	02-02-16/13:20:05
52	60.0	F	02-02-16/13:20:05
53	61.7	F	02-02-16/13:20:06
54	61.7	F	02-02-16/13:20:06
55	62.2	F	02-02-16/13:20:06
56	60.0	F	02-02-16/13:20:07
57	58.4	F	02-02-16/13:20:07
58	58.4	F	02-02-16/13:20:08
59	59.5	F	02-02-16/13:20:08
60	59.5	F	02-02-16/13:20:09
61	59.9	F	02-02-16/13:20:09
62	59.9	F	02-02-16/13:20:10
63	54.6	F	02-02-16/13:20:10
64	52.1	F	02-02-16/13:20:10
65	53.4	F	02-02-16/13:20:11
66	53.4	F	02-02-16/13:20:11
67	53.4	F	02-02-16/13:20:12
68	52.1	F	02-02-16/13:20:12
69	53.7	F	02-02-16/13:20:13
70	55.2	F	02-02-16/13:20:13
71	55.5	F	02-02-16/13:20:14
72	55.5	F	02-02-16/13:20:14
73	56.3	F	02-02-16/13:20:14
74	56.8	F	02-02-16/13:20:15
75	56.3	F	02-02-16/13:20:15
76	55.9	F	02-02-16/13:20:16
77	46.5	F	02-02-16/13:20:16
78	45.5	F	02-02-16/13:20:17
79	48.7	F	02-02-16/13:20:17
80	46.5	F	02-02-16/13:20:18
81	46.5	F	02-02-16/13:20:18
82	46.5	F	02-02-16/13:20:18
83	**46.5**	F	02-02-16/13:20:19
84	**47.1**	F	02-02-16/13:20:19
85	**54.6**	F	02-02-16/13:20:20
86	**55.2**	F	02-02-16/13:20:20
87	53.9	F	02-02-16/13:20:21
88	53.9	F	02-02-16/13:20:21
89	53.9	F	02-02-16/13:20:22

Figure 9

NO (Reading)	Zone	Zone Description
1 – 25	Warm	Coronary band to crest of capsule.
25 – 28	Warm → Cool	Crest of capsule to lower shoulder of coronary groove.
28	Cool	☉
28 – 84	Cool	☉ to inner hoof wall's interface with lamellar and solar coriums.
84 – 86	Warm	Interface with the lamellar and solar coriums.
86 – 89	Warm → Cool	Slight cooling to terminal end of toe wall.

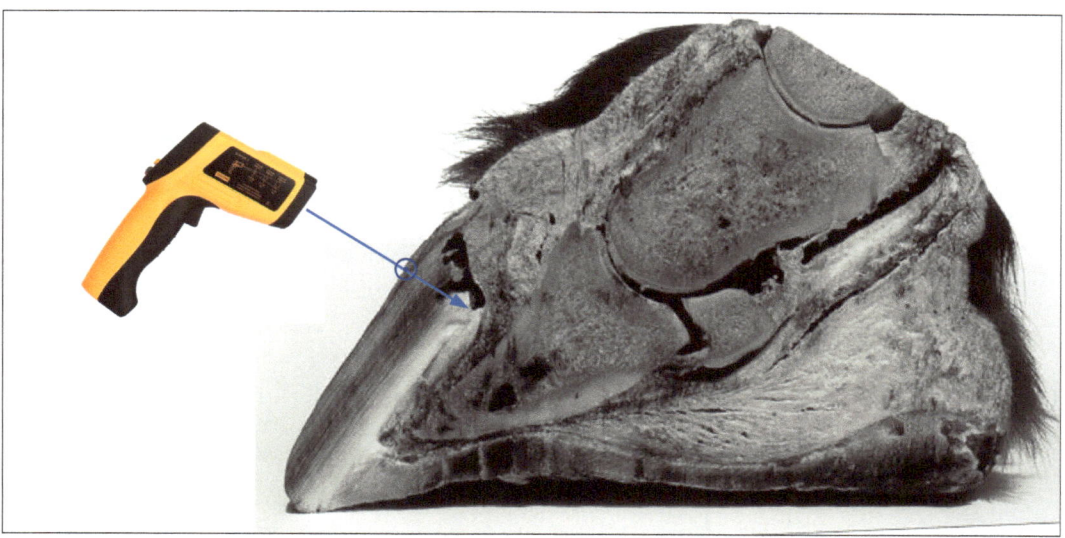

Figure 11. Aiming the laser pointer gun.

(Continued from page 7)

Conclusion

The key finding of this study shows unequivocally a distinct and rapid descent in temperature from the crest of the capsule down to the lower shoulder of the coronary groove. It is at this lower temperature reading that the full width of the *straum medium* begins and where ☉ is to be marked on the MATW.

Using the Laser Gun

- The laser gun is aimed at approximately 90° to the MATW (*Figure 11*).

- Scan the entire length of the MATW beginning at the coronary band. Repeat scan 3 times to confirm consistency of data. If using a bluetooth connection to a laptop, an assistant should alert the scanning person ("scanner") when ☉ occurs. Narrow scans to this region of the MATW and repeat scans again to confirm ☉.

- If using a simple laser gun without bluetooth connectivity, take multiple readings in close proximity of each other until ☉ is confirmed; the scanner or an assistant should mark ☉ accordingly.

Measuring H° and H°TL

In practice, ⊙ is drawn as a crosshair (⊕) on the MATW as explained in the *Basic Guidelines* of the Natural Trim (*Figure 12*).[1] To locate H°, a secondary line (—) is then drawn 1 cm below ⊕. When the hoof is viewed in its lateral profile, this 1 cm long linear axis is measurable as H° on the HMR (*Figure 13*). H°TL is read from ⊕ to SP using the HMR (*Figure 14*).

Measuring B° and B°TL

If H° is not readable with the HMR due to DTAs,[2] the *Advanced Guidelines* of the Natural Trim are used.[3]

---

[1] Jaime Jackson. *The Natural Trim: Basic Guidelines.* (2019)
[2] "Divergent toe wall angle" (DTA) is any change or bend in the angle of the MATW away from H°TL; DTAs are commonly associated with laminitis and invasive rasping of the toe wall.
[3] Jaime Jackson. *The Natural Trim: Advanced Guidelines.* (2019)

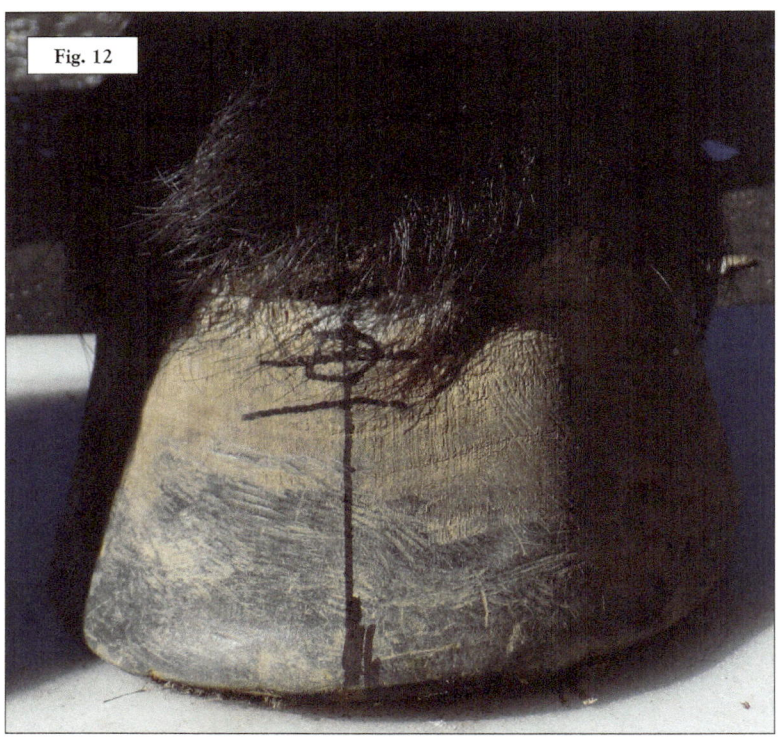

Figure 12. ⊕ is located with the laser gun, then drawn on the MATW with a black Sharpie. To find H°, a second line is draw 1 cm below also on the MATW. [Note: Gridlines and ⊕ can be removed with the HB-1 buffer/sander or common sandpaper.]

Measuring H° and H°TL · Measuring B° and B°TL

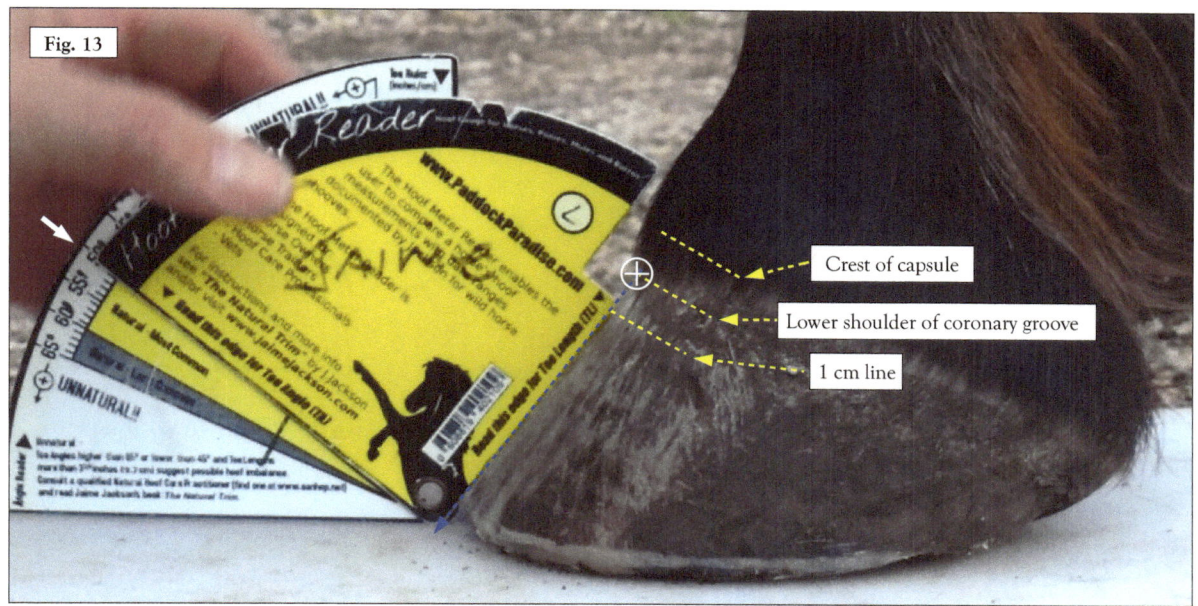

Figure 13. The Toe Angle Reader of the HMR is set parallel with the MATW between ⊕ and the lower 1 cm line. H° here measures 51°.

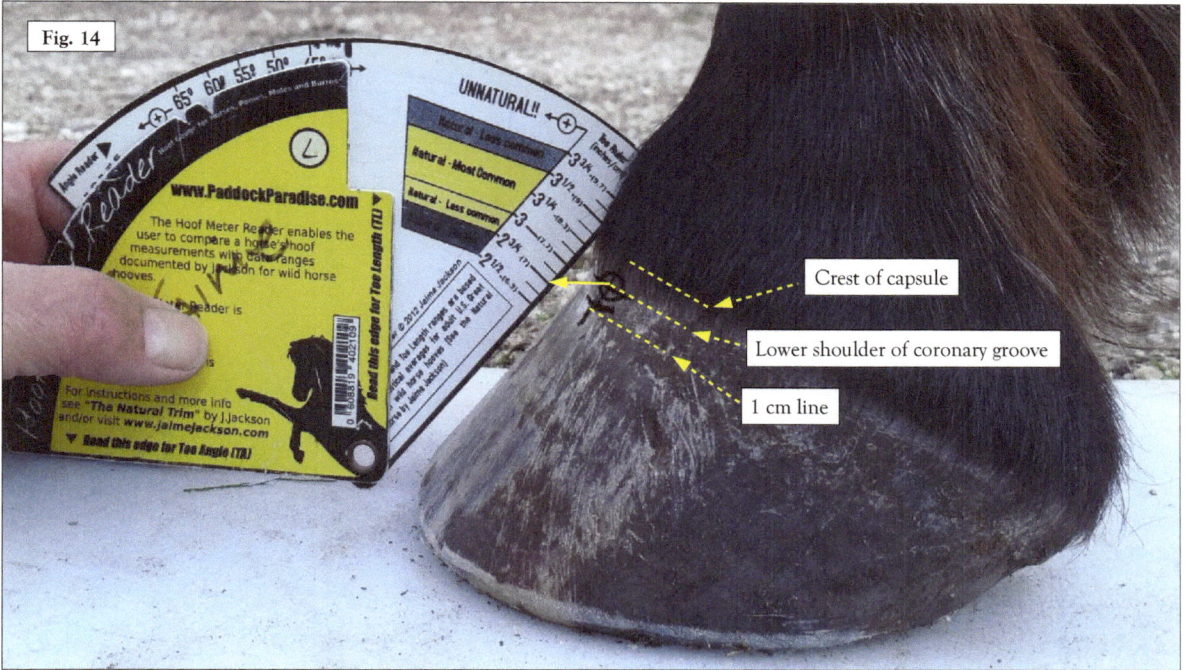

Figure 14. The Toe Ruler dial of the HMR is set parallel to the MATW between ⊕ and the lower 1 cm line. The Toe Ruler is read straight across from ⊕ parallel to SP. H°TL here measures 2¾ in (7 cm).

Impact of ambient temperatures on the location of ⊕

During the Lompoc study, a significant descent in the ambient air temperature occurred that dramatically depressed the readings taken with the laser gun. However, this did not affect the relative temperature differentials defining ⊕. The question may arise: Why does the surface termperature of the MATW change with changes in ambient air termperature?

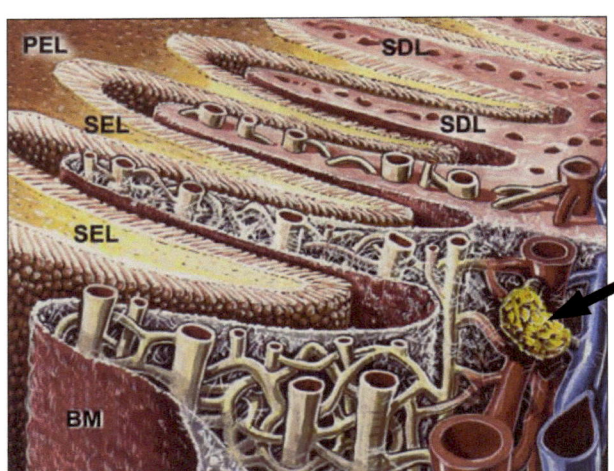

Figure 15. *Black arrow* points to artery-vein shunt (anastomosis) within the lamellar Supercorium. P3 is off to the right (not shown); the *s. internum* is off to the left (not shown). [PEL: primary epidermal laminae; SEL: secondary epidermal laminae; BM: basement membrane; SDL: secondary dermal laminae.]

According to Professor Christopher Pollitt of the Australian Laminitis Research Unit at Queensland University in Australia, a network of arteriovenous anastamoses ("AVAs" or AV shunts) regulate blood flow between the arteries and veins of the Supercorium (Figure 15).

When the AV shunts open, blood circulates more internally, that is, away from the epidermal armor. Hence, we see cooler termperature readings down the MATW with the laser gun. The intent of nature here is to regulate, in this instance warm, the foot within, thereby facilitating homeostasis. This is what happened with Apollo as a blanket of cold marine air flowed over our Paddock Paradise.

Had it been the opposite, with a marine layer giving way to warm or hot sunny skies, the AV shunts would close, and blood could then flow more freely towards the inner hoof wall, including at ⊕. The effect would then be to cool the foot within but warm the epidermal armor.

Looking ahead to the future, new software would be welcomed that could standardize thermal readings for ⊕ relative to flucuations in ambient temperature changes. Further research would be needed to see if these thermal readings were the same for all horses, or, if not, NHC practitioners could create individualized thermal charts for horses in their care. These charts could be extremely valuable in detecting subtle and acute thermal deviations in the Supercorium due to laminitis (WHID).

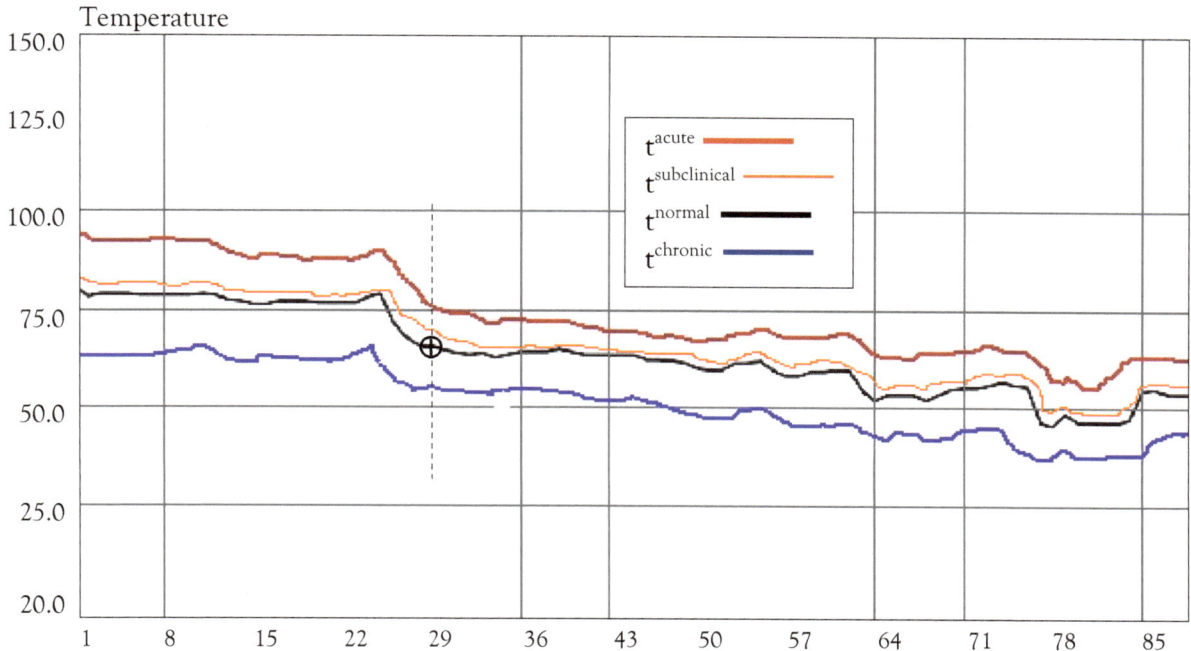

Figure 15: Hypothetical fluctuations in thermal readings due to laminitis.

Impact of laminitis on thermal fluctuations along the MATW

Since the Lompoc study, I've asked myself would thermal readings along the MATW be useful in the diagnosis of laminitis in its various developmental stages? Conversely, could these readings also be useful in monitoring anti-inflammatory interventions within the Supercorium?

We might predict that in the early *subclinical stage* laser readings would show a low but measureable elevation in termperature along the MATW due to inflammation of the Supercorium. As the *clinical stage* arrives then we should expect a dramatic thermal elevation. When the *chronic stage* arrives with the emergence of DTAs and B°TL, we might expect, in the absence of NHC intervention, ⊕ → SP to become a cool zone as lamellar wedge emeges during the separation of the inner hoof wall from P3. However, it is feasible that the volar dome could be scanned along the MAVP to detect the location of the lamellar dermis during P3 rotation. Figure 15 provides hypothetical changes along the MATW due to laminitis.

I would add also, from observation, that when there is catastrophic shedding of the entire, or a portion of, the hoof wall from the foot, the epidermal basal cells (still attached to the basal membrane and dermal laminae) are capable of generating a dense, tough layer of secondary epidermal laminae that can dessicate and seal off P3. I suspect the face of P3 then also becomes a cool zone, but targeted research would be necessary to confirm this. My experience has been that this layer is actually dormant and capable of regenerating new and healthy attachments to newly produced primary epidermal laminae descending when there has been NHC intervention.

Definitions

Bull's-eye (◉). A single point on the MATW, through which an axis set at 90° to the MATW passes through the lower shoulder of the coronary groove, marking the beginning of the full thickness of the hoof wall.

Critical Measurements. Measurements derived from the Navigational Landmarks used by the NHC Practitioner to safely and accurately conduct the Natural Trim.

Four Pillars of Natural Horse Care (aka, 4 Pillars of NHC). Humane horse care based on the wild, free-roaming horse model of the U.S. Great Basin: Paddock Paradise (tracking system), a reasonably natural diet, natural horsemanship based on the specie's Natural Gait Complex, and the Natural Trim.

Herd Management Area (MHA). Are public lands under the supervision of the United States Bureau of Land Management (BLM) that are managed for the primary but not exclusive benefit of wild, free-roaming horses and burros.

H° (Healing Angle). A specific measurement of the hoof's angle of growth along the MATW used by NHC Practitioners when conducting the Natural Trim. Applies only to horses living in human captivity.

H°TL (Healing Angle Toe Length. A specific measurement of the hoof's toe length along the MATW used by NHC Practitioners when conducting the Natural Trim. Applies only to horses living in human captivity.

H° and H°TL. Critical Measurements statistically derived from their wild, free-roaming horse counterparts of the U.S. Great Basin, N° and N°TL.

HMA. (*See* Herd Management Area).

Homeostatis. The maintenance of relatively stable internal physiological conditions (as body temperature or the pH of blood) in higher animals under fluctuating environmental conditions.

MATW-MAVP Joint. The point at which the MAVP ends and the MATW begins.

Median Axis of the Toe Wall (MATW). A line drawn from the MATW-MAVP Joint extending vertically to the crest of the capsule following the stratification of tubular horn visible in the outer hoof wall.

Median Axis of the Volar Plane (MAVP). A line drawn across the entire bottom of the hoof passing through the cleft of the heel bulbs, the central frog sulcus, the point of frog, terminating at its intersection with the outer wall of the *stratum medium*. The MAVP establishes the point of origin of the MATW, and, thus, is the basis for the MATW-MAVP Joint.

Natural Gait Complex. The natural gaits and their variations seen among America's wild, free-roaming horses living HMAs in the U.S. Great Basin under federal government protection. Such natural movement is rooted in the ancient behavioral biology of their species, *Equus ferus ferus*. It is extant in all horses living today, although

Definitions

corrupted in horses living in captivity due to human meddling.

Natural Hoof Care. The Natural Trim conducted within the holistic framework of the 4 Pillars of Natural Horse Care.

N° (Natural Toe Angle). A specific measurement of the wild horse hoof's angle of growth along the MATW. Applies only to wild, free-roaming horses living 24/7 in U.S. Great Basin HMAs.

N°TL (Natural Toe Length). A specific measurement of the wild horse hoof's toe length along the MATW. Applies only to wild, free-roaming horses living 24/7 in U.S. Great Basin HMAs.

Natural Trim. A humane trim method that mimics the natural wear patterns of America's wild, free-roaming horses living 24/7 in U.S. Great Basin HMAs.

Natural Trim Hoof Plexus. A three-dimensional model of intersecting lines and planes that define the griding system used in the Natural Trim. The Navigational Landmarks and Critical Measurements are derived from this model.

Navigational Landmarks. A series of gridlines derived from the Natural Trim Hoof Plexus marked on the hoof that enable the NHC Practitioner to make Critical Measurements.

Supercorium. The interconnecting and intercommunicating network of coriums responsible for creating, nutrifying, and replacing the hoof capsule.

Support Plane. A plane representing a flat surface that supports the hoof and defines its VP.

Theory of H° (Healing Angle). A set of guiding principles formulated into a Natural Trim methodology derived from the wild horse model that helps to explain or predict certain healing and pathological phenonmena occurring across the horse's body based on specific thermal and mass changes occurring in the horse's hooves.

Volar Plane. Imaginary plane formed by the active wear of the hoof that are defined by the Support Plane.

Text References

Equine Laminitis – Current Concepts. Christopher C. Pollitt. (May 2008: Rural Industries Research and Development Corporation, Australian Government). RIRDC Publication No 08/062. RIRDC UQ-118A.

The Natural Horse: Lessons From the Wild. Jaime Jackson (2020 revised edition).

The Natural Trim: Advanced Guidelines. Jaime Jackson (2019).

The Natural Trim: Basic Guidelines –Trim Mechanics, Biodynamics, and healing Forces in Paddock Paradise: Working with Naturre to Create the Pefect Hoof. Jaime Jackson. (2019)

The Natural Trim: Principles and Practice. Jaime Jackson (2012; revised 2019).

Image Attributions

Cover (front)
- Jaime Jackson

Cover (back)
- *Top L/R*: Jaime Jackson
- *Center*: Jill Willis
- *Lower*: Jaime Jackson archives

Title page
- Jaime Jackson

P. 1
- Jaime Jackson
- Facing page: Jaime Jackson

P. 2
- Left: Jaime Jackson
- Right: Jill Willis

P. 3
- Jaime Jackson

P. 4
- Jaime Jackson

P. 7
- https://commons.wikimedia.org/wiki/File:1024_Pyrometer-8445.jpg

P. 6
Jaime Jackson

P. 8
- Jaime Jackson

P. 9-10
- Jaime Jackson

P. 11
- Jaime Jackson

P. 12
- Jill Willis

P. 13
- Jill Willis

P. 14
- John McDougall/Chris Pollitt

P. 15
- Jaime Jackson

Study Technology

- WCI Professional High-Temperature IR Thermometer Laser Pointer Gun with Wireless USB Computer Interface and Type K Probe - Instant °C or °F Measurement with Calibrated Graphics Software.
- Hoof Meter Reader: https://jaimejackson.com/collections/measuring-and-balancing-tools/products/hoof-meter-reader

NHC Links

www.ISNHCP.net
Institute for the Study of
Natural Horse Care Practices
Founded 2009
International Training Program
for Professional Natural Hoof Care Practitioners
ISNHCP Veterinary Consultation Services

www.AANHCP.NET
Association for the Advancement of
Natural Horse Care Practices
Founded in 2000; 501c3 in 2004
Advocacy organization for humane horse care practices
based on the U.S. Great Basin Wild, Free-Roaming Horse Model,
Listings of NHC Practitioners, NHC Bulletins and Articles

www.PaddockParadise.net
International website promoting natural boarding for horses
based on the concept of Paddock Paradise by Jaime Jackson

NHC Facebook Pages
AANHCP, ISNHCP, Paddock Paradise, The Natural Trim

www.JaimeJackson.com
Author's website promoting NHC
Books, Tools/Equipment, Consultations

www.ingramcontent.com/pod-product-compliance
Lightning Source LLC
Chambersburg PA
CBHW041108070526
44583CB00002B/116